Interactive
Mathematics Program®

INTEGRATED HIGH SCHOOL MATHEMATICS

Pennant Fever

FIRST EDITION AUTHORS:
Dan Fendel, Diane Resek, Lynne Alper, and Sherry Fraser

CONTRIBUTORS TO THE SECOND EDITION:
Sherry Fraser, IMP for the 21st Century
Jean Klanica, IMP for the 21st Century
Brian Lawler, California State University San Marcos
Eric Robinson, Ithaca College, NY
Lew Romagnano, Metropolitan State College of Denver, CO
Rick Marks, Sonoma State University, CA
Dan Brutlag, Meaningful Mathematics
Alan Olds, Colorado Writing Project
Mike Bryant, Santa Maria High School, CA
Jeri P. Philbrick, Oxnard High School, CA
Lori Green, Lincoln High School, CA
Matt Bremer, Berkeley High School, CA
Margaret DeArmond, Kern High School District, CA

Key Curriculum Press

Second Edition   I M P

This material is based upon work supported by the National Science Foundation under award numbers ESI-9255262, ESI-0137805, and ESI-0627821. Any opinions, findings, and conclusions or recommendations expressed in this publication are those of the authors and do not necessarily reflect the views of the National Science Foundation.

Key Curriculum Press
1150 65th Street
Emeryville, California 94608
email: editorial@keypress.com
www.keypress.com
10 9 8 7 6 5 4 3 2 1   14 13 12 11
ISBN 978-1-60440-050-2
Printed in the United States of America

Project Editors
Mali Apple, Josephine Noah, Sharon Taylor

Project Administrators
Emily Reed, Juliana Tringali

Professional Reviewers
Rick Marks, Sonoma State University, CA
D. Michael Bryant, Santa Maria High School, CA, retired

Accuracy Checker
Carrie Gongaware

First Edition Teacher Reviewers
Daniel R. Bennett, Moloka'i High School, HI
Maureen Burkhart, Northridge Academy High School, CA
Dwight Fuller, Ponderosa High School, CA
Daniel S. Johnson, Silver Creek High School, CA
Brian Lawler, California State University San Marcos, CA
Brent McClain, Vernonia School District, OR
Susan Miller, St. Francis of Assisi Parish School, PA
Amy C. Roszak, Cottage Grove High School, OR
Carmen C. Rubino, Silver Creek High School, CA
Barbara Schallau, East Side Union High School District, CA
Kathleen H. Spivack, Wilbur Cross High School, CT
Wendy Tokumine, Farrington High School, HI

First Edition Multicultural Reviewers
Genevieve Lau, Ph.D., Skyline College, CA
Arthur Ramirez, Ph.D., Sonoma State University, CA
Marilyn Strutchens, Ph.D., Auburn University, AL

Copyeditor
Brandy Vickers

Interior Designer
Marilyn Perry

Production Editor
Andrew Jones

Production Director
Christine Osborne

Editorial Production Supervisor
Kristin Ferraioli

Compositor
Lapiz Digital Services, Kristin Ferraioli

Art Editor/Photo Researcher
Maya Melenchuk

Technical Artists
Lapiz Digital Services, Laurel Technical Services, Maya Melenchuk

Illustrators
Taylor Bruce, Deborah Drummond, Tom Fowler, Briana Miller, Evangelia Philippidis, Sara Swan, Diane Varner, Martha Weston, April Goodman Willy

Cover Designer
Jeff Williams

Printer
Lightning Source, Inc.

Mathematics Product Manager
Elizabeth DeCarli

Executive Editor
Josephine Noah

Publisher
Steven Rasmussen

SUPPLEMENTAL ACTIVITIES

Pennant Fever—Permutations, Combinations, and the Binomial Distribution

Pennant Fever

Permutations, Combinations, and the Binomial Distribution

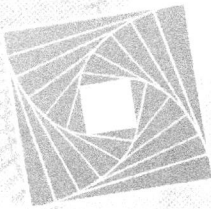

Pennant Fever—Permutations, Combinations, and the Binomial Distribution

Play Ball!

In this unit, the central problem involves two baseball teams in a pennant race. Your main task will be to find each team's probability of winning the pennant.

In the first few activities, you will have a chance to speculate about the problem. Over the course of the unit, you will occasionally leave the central problem to explore a variety of other situations, including the chance of finding two people in a room with the same birthday.

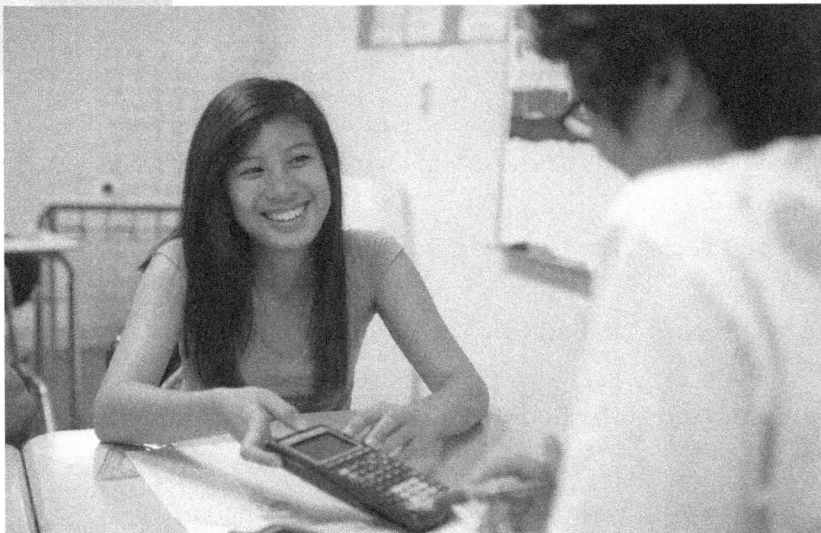

Gabi Reyes and Brandon Yi determine the baseball teams' probabilities.

Race for the Pennant!

It's almost the end of the baseball season, and only two teams still have a chance to win the pennant: the Good Guys and the Bad Guys. Here are their records for the season so far.

Team	Games won	Games lost	Games left
Good Guys	96	59	7
Bad Guys	93	62	7

The Good Guys and the Bad Guys will not play against each other in any of their remaining games.

The central problem of this unit is to find the probability that the Good Guys will win the pennant.

1. Study the possibilities for each team.

   a. What is the best record the Good Guys could have at the end of the season? That is, what is the most wins and fewest losses they could end up with?

   b. What is the worst record the Good Guys could have at the end of the season?

   c. What is the best record the Bad Guys could have at the end of the season?

   d. What is the worst record the Bad Guys could have at the end of the season?

2. a. Discuss with your group what you think is the most likely outcome for each team in its remaining seven games.

   b. Make your own decision about part a, and give reasons to support your conclusion.

continued ⬥

3. a. Discuss with your group the Good Guys' probability of winning the pennant. Try to come to agreement on the likelihood that they will win.

   b. Give your best guess right now of this probability, and explain your thinking. Be sure to state any assumptions you make.

Happy Birthday!

The seven-day week is used throughout the world. Many cultures have sayings about how the day of the week on which a person is born might affect that individual's personality.

People sometimes use as role models individuals born on the same day of the week they were born. Here are some famous people born on each day of the week

Sunday: Louis Armstrong (Aug. 4, 1901)
Whoopi Goldberg (Nov. 13, 1955)
Katharine Hepburn (May 12, 1907)
Michael Jordan (Feb. 17, 1963)
Elizabeth Cady Stanton (Nov. 12, 1815)

Monday: Henry "Hank" Aaron (Feb. 5, 1934)
Michelangelo Buonarroti (Mar. 6, 1475)
Bill Clinton (Aug. 19, 1946)
Natalie Coughlin (Aug. 23, 1982)
Jodie Foster (Nov. 19, 1962)

Tuesday: Dr. Martin Luther King Jr. (Jan. 15, 1929)
Mao Tse-tung (Dec. 26, 1893)
Golda Meir (May 3, 1898)
Michelle Pfeiffer (Apr. 29, 1958)
Tiger Woods (Dec. 30, 1975)

Wednesday: Maya Angelou (Apr. 4, 1928)
Roald Dahl (Sep. 13, 1916)
Dr. Seuss (Mar. 2, 1904)
Steven Spielberg (Dec. 18, 1946)
Desmond Tutu (Oct. 7, 1931)

Thursday: Louisa May Alcott (Nov. 29, 1832)
Marie Curie (Nov. 7, 1867)
Nelson Mandela (July 18, 1918)
John Steinbeck (Feb. 27, 1902)
Laura Ingalls Wilder (Feb. 7, 1867)

continued

Friday:	Madeleine L'Engle (Nov. 29, 1918)
	Michelle Obama (Jan. 17, 1964)
	Jesse Owens (Sep. 12, 1913)
	Mother Teresa (Aug. 26, 1910)
	Oprah Winfrey (Jan. 29, 1954)
Saturday:	Amelia Earhart (July 24, 1897)
	Mahatma Gandhi (Oct. 2, 1869)
	Eleanor Roosevelt (Oct. 11, 1884)
	J. K. Rowling (July 31, 1965)
	Booker T. Washington (Apr. 5, 1856)

This raises an interesting question:

Do you know what day of the week you were born on? How would you figure it out if you didn't know?

Your POW task is to develop a system for determining the day of the week on which someone was born, based on the date that person was born.

1. Use only a calendar for the current month and the "Basic Information About Calendars" provided here to figure out on which day of the week you were born. Do not look at calendars for any other month of this year or for other years.

2. Develop general directions so that someone else could apply the method you used in Question 1 to determine the day of the week on which he or she was born.

3. Have someone try to use the directions you created and then tell you how well your directions worked. Based on this feedback, make changes to correct or clarify your directions.

continued

Basic Information About Calendars

POW
10

The number of days in each month, except February, is the same every year. February gets an extra day in leap years. For the period from 1901 through 1999, leap years are those years that are multiples of 4: 1904, 1908, 1912, and so on. (For this POW, only consider birth dates from 1901 through 1999. There are special rules for years that are multiples of 100.)

Here is a list of the number of days in each month.

- January: 31 days
- February: 28 days (but 29 in leap years)
- March: 31 days
- April: 30 days
- May: 31 days
- June: 30 days
- July: 31 days
- August: 31 days
- September: 30 days
- October: 31 days
- November: 30 days
- December: 31 days

Write-up

1. *Problem Statement*

2. *Process*

3. *Solution:* Describe how you determined the day of the week on which you were born, and give the general directions you created. Also write about the experience of having someone else use your directions. If you modified the directions based on that person's feedback, describe the changes you made.

4. *Self-assessment*

Playing with Probabilities

1. Suppose someone throws a dart at this target. Assume the dart will hit the target, with all points equally likely to be hit.

   What is the probability that a given dart will land in the black area? In the white area? In the shaded area? Explain your answers.

2. Draw an area model that represents a situation with three outcomes. One outcome should have a probability of $\frac{1}{5}$, the second should have a probability of .7, and the third should have a probability of—well, you figure out the probability of the third outcome.

   Explain how you found the third probability and how your area model represents these probabilities.

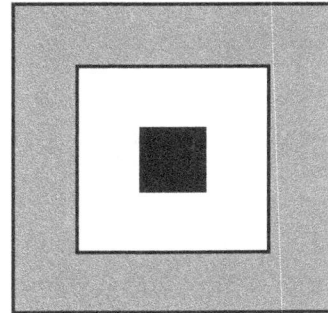

3. Make up a situation with three outcomes that would have the probabilities in Question 2.

4. Ms. Hernandez and her twins are in front of a gumball machine that contains red, blue, and purple gumballs. Nine gumballs are purple, 25% of the gumballs are red, and the remaining 60% are blue.

   How many of the gumballs are red and how many are blue? Explain how you know your answer is correct.

Special Days

This activity will help you get started on your POW. It asks you to determine what day of the week three special events fell on during the calendar year *previous* to the current year.

Remember to take into account whether either the current year or the previous year was a leap year.

Explain each of your answers. As in the POW, determine these days of the week without consulting any calendar except one for the current month.

1. New Year's Day (January 1)
2. Valentine's Day (February 14)
3. Your birthday

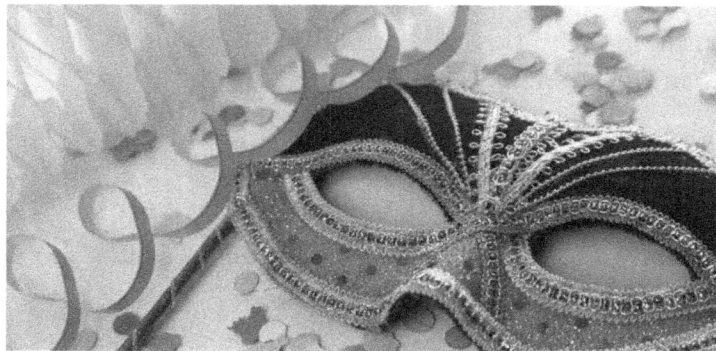

Trees and Baseball

One of the best techniques for analyzing situations like the baseball problem is the tree diagram. You may recall using tree diagrams in the Year 1 unit *The Game of Pig*.

Over the next several days, you'll apply this technique to several situations. You'll also get your first definitive probability results for particular possible outcomes of the baseball pennant race.

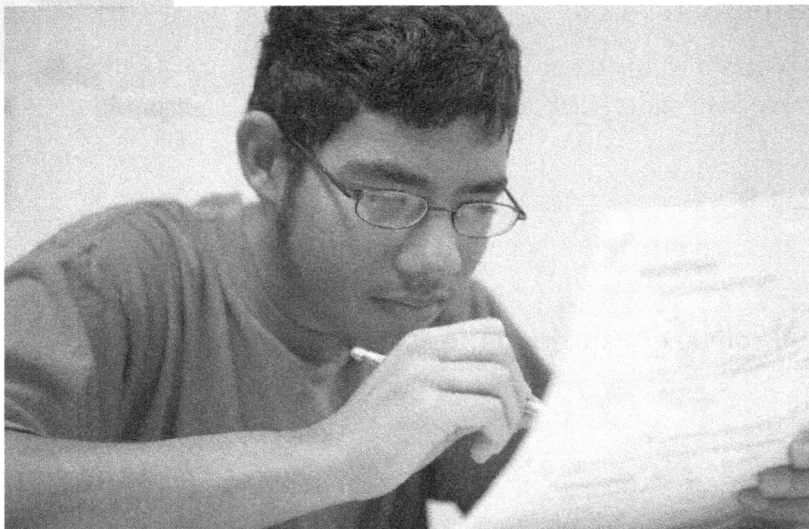

Alexander Reyes uses a tree diagram to help solve a probability problem.

Choosing for Chores

Part I: Wash or Dry?

Scott and Letitia are brother and sister. After dinner, they have to do the dishes, with one washing and the other drying. They are having trouble deciding who will do which task, so they come up with a method based on probability.

Letitia grabs some spoons and puts them in a bag. Some have purple handles and others have green handles. Scott has to pick two of the spoons. If their handles are the same color, Scott will wash. If they are different colors, he will dry.

It turns out that there are two purple spoons and three green ones. What is the probability that Scott will wash the dishes? Explain your answer.

Part II: Allowance Choices

Scott and Letitia's parents want to encourage them to learn more about probability. They gave Scott and Letitia two choices for how to get paid for their chores.

- **Choice 1:** They get $4.
- **Choice 2:** They pick two bills out of a bag that contains four $1 bills and one $5 bill.

Which choice would you take? Explain your reasoning carefully.

Baseball Probabilities

Willie is one of the star hitters for the Good Guys. Every time he comes up to bat, he has one chance in three of getting a hit. In standard baseball terminology, we would say his batting average is .333.

1. Suppose Willie comes up to bat twice in a certain game.
   a. What is the probability that he'll get a hit both times?
   b. What is the probability that he won't get a hit either time?
   c. Use your answers to find the probability that he will get exactly one hit.

2. Suppose that in another game, Willie comes up to bat three times.
   a. What is the probability that he'll get a hit all three times?
   b. What is the probability that he won't get any hits?
   c. Use your answers to find the probability that the number of hits will be either 1 or 2.

Possible Outcomes

The Good Guys and the Bad Guys can each achieve many different records for their final seven games. For instance, they could both win the rest of their games, or one team could win the rest of its games while the other team loses all of its games.

1. List the possible records the Good Guys could have for their final seven games.

2. List the possible records the Bad Guys could have for their final seven games.

3. a. How many combinations of records are there for the two teams? For instance, one combination is that the Good Guys win six games and lose one while the Bad Guys win three games and lose four.

   b. Make a table or other display showing all the possible combinations of records. Indicate in your display which team ends up winning the pennant in each case.

How Likely Is "All Wins"?

For the rest of this unit, use .62 as the probability that the Good Guys will win any given game. Use .6 as the probability that the Bad Guys will win any given game. These values come from the teams' current percentages of winning games. Remember that the two teams will not play against each other.

1. Find the probability that the Good Guys will win all seven of their remaining games. Justify your conclusion with a tree diagram or area model.

2. Find the probability that the Bad Guys will win all seven of their remaining games.

3. What is the probability that both teams will finish the season this way? That is, find the probability that both the Good Guys and the Bad Guys will win all seven of their remaining games.

Go for the Gold!

This is your lucky week. You received an invitation to be a contestant on the television game show *Go for the Gold!* You'll have a chance to win a lot of money if you accept the invitation. However, the *Go for the Gold!* producers require a nonrefundable fee of $100 for the opportunity to play.

Here's how the game works.

> You are given a jar containing two white cubes and one gold cube. You choose a cube without looking (so each cube is an equally likely choice). If you get the gold cube, you are shown another jar. If you don't get the gold cube, you are out.

> The second jar contains four white cubes and one gold cube. Again, you choose one cube from the jar without looking. If you get the gold cube, you win $1,000. If not, you get nothing.

1. a. What is the probability that you will pick the gold cubes from both jars and win the $1,000 prize? Use both an area model and a tree diagram to justify your results.

   b. Pick a section of your area model from part a, and explain what path of your tree diagram it corresponds to.

2. If you win the game (by picking the two gold cubes), you get $1,000. But you have to pay $100 just to play the game.

   Do you accept the show's offer to be a contestant? Explain your decision.

Diagrams, Baseball, and Losing 'em All

1. a. Compare the process of calculating probabilities using area models and tree diagrams. Describe the advantages and disadvantages of each method. Use specific examples to illustrate your ideas.

   b. When might you be better off using no diagram at all?

2. Where do you stand on solving the unit problem? That is, what parts of the problem have you solved and what remains to be figured out?

3. What is the probability that the season will end up with both the Good Guys and the Bad Guys losing all seven of their remaining games?

The Birthday Problem

In the first POW of this unit, you developed a method for finding the day of the week on which someone was born. Beginning with *Day-of-the-Week Matches,* you start a sequence of activities that continue this birthday theme. You may find the answer to the final activity in this sequence quite surprising.

First, though, you will work on an activity that relates to the new POW.

Sheila Roberts explains to her classmates what the probability is that there would be at least one birthday match in a group the size of their class.

Let's Make a Deal

Congratulations! You've been selected to be a contestant on the *Let's Make a Deal* television show! Let's review how the game works.

> You are shown three doors, labeled A, B, and C. Behind one of the doors is a brand new sports car. Behind the other two doors are worthless prizes.

> You select a door. The game-show host then opens one of the other doors. Because he knows where the car is hidden, he makes sure to open a door that reveals a worthless prize.

> You now have a choice. You can either stay with the door you originally selected or switch to the remaining closed door.

> You will win whatever is behind the door you choose this time.

Here's the big question for you (and for this POW).

> *Are your chances of winning the car better if you stay with your original choice, or are they better if you switch to the remaining closed door, or are they equally likely with the two strategies?*

More precisely, your task is to find the probability of getting the car for each of the two strategies. Write a careful explanation of how you found these probabilities.

continued

○ *Write-up*

1. *Problem Statement*

2. *Process:* Include a description of how you and your partner did the simulation in the activity *Simulate a Deal.* Give the results of your simulation, and discuss how that activity contributed to your understanding of the problem.

3. *Solution:* Give the probability of winning for each strategy, including an explanation of how you found the probabilities.

4. *Self-assessment*

Adapted with permission from *Mathematics Teacher,* © April 1991 by the National Council of Teachers of Mathematics.

Simulate a Deal

In the POW *Let's Make a Deal,* you need to decide whether to switch to the remaining closed door or stay with your original choice.

In this activity, you and a partner will simulate the problem. That is, you will use each of the two strategies (switch or stay) a number of times and find out what happens.

1. First try the strategy in which the contestant always switches his or her guess after being shown the open door. One of you will play the game-show host and the other will play the contestant. Come up with some way to realistically act out the situation.

   Play the game ten times using this strategy, keeping track of how many times the contestant wins.

2. Switch roles with your partner and try the strategy in which the contestant always stays with the original door. Play the game ten times using this strategy, keeping track of how many times the contestant wins.

Day-of-the-Week Matches

The POW *Happy Birthday!* involves the day of the week on which a person is born. This activity is the first of several that connect that theme to probability.

In this activity, assume that each of the seven possibilities—Sunday through Saturday—is equally likely.

1. Imagine a random group of people. Is there a group size for which you would be certain that at least two people in the group were born on the same day of the week? What's the smallest such group size? Explain your answer.

2. If you pick two people at random, what is the probability that they were born on different days of the week? What is the probability that they were born on the same day of the week? Justify your answers.

3. If you pick three people at random, what is the probability that all three were born on different days of the week? What is the probability that at least two of the three people were born on the same day of the week? Justify your answers.

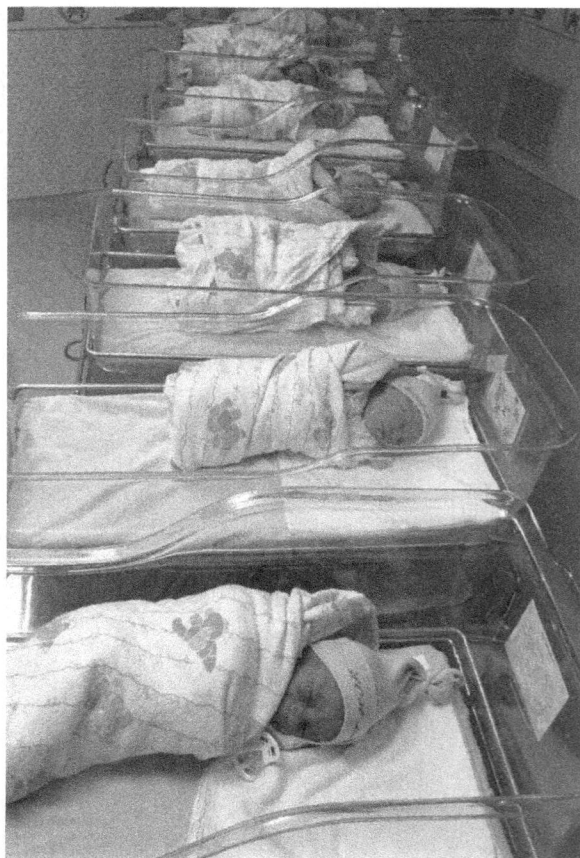

Day-of-the-Week Matches, Continued

In *Day-of-the-Week Matches,* you found the probability that two people chosen at random were born on the same day of the week and then the probability for three people.

You also found that if there were at least eight people, the probability would be 1, because you would be certain of a match.

Now you will investigate this question.

> *What is the minimum number of people needed so that the probability of having at least one day-of-the-week match is greater than $\frac{1}{2}$?*

As before, assume each day of the week is equally likely.

Be sure to explain your answer. (Because the probability for two people and the probability for three people are both less than $\frac{1}{2}$, you know that you need more than three people. You also know that the number can't be more than eight. It may be easier to figure out the probability that a group of people were all born on *different* days of the week than to figure out the probability that at least two were born on the *same* day of the week.)

Monthly Matches

You've been working on problems involving the probability that two people are born on the same day of the week. Now you move from days of the week to months of the year.

Although the months are not all the same length, make the simplifying assumption in this activity that each month is equally likely to be the birth month for a person chosen at random.

1. What is the smallest number of people needed for the probability to be 100% that at least two of them were born in the same month?

2. If you pick a person at random, what is the probability that you and that person were born in the same month?

3. If you have a group of three people chosen at random, what is the probability that at least two of them have the same birth month? What is the probability for a four-person group?

4. How many people would you need in a randomly chosen group for the probability of a month-of-birth match in the group to be greater than $\frac{1}{2}$?

The Real Birthday Problem

You've been working on day-of-the-week matches and month-of-the-year matches. Now you're ready to solve a very famous problem known simply as the "birthday problem."

What is the minimum number of people needed so that the probability of having at least one birthday match is greater than $\frac{1}{2}$?

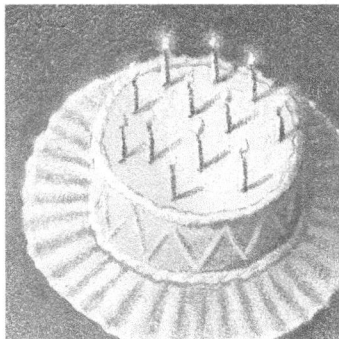

For this problem, ignore leap years. That is, assume that a year has 365 days. Also assume that each of these 365 days is equally likely to be a person's birthday.

1. Before doing any computation or analysis, make a guess about the answer to the birthday problem and write it down.

2. Now do the necessary analysis to find the answer, and explain your work.

Six for the Defense

Mari wants to be a defense attorney. She is very excited because her civics teacher has just announced that the class will soon begin a four-day unit on the court system.

The class will act out a different famous court case each day. At the beginning of each class, the teacher will use a die to decide each student's role. If the die comes up 6, the student will be one of the defense attorneys for that case. If the die comes up 1 through 5, the student will play some other role.

1. Make a tree diagram or area model for the situation that will help Mari analyze how often she is likely to be a defense attorney over the four days.

2. What is the probability that Mari will be a defense attorney every day of the unit?

3. What is the probability that Mari will never be a defense attorney during the unit?

4. On the third day of the unit, the class will act out a case with which Mari is very familiar. What she'd like best is to be a defense attorney on that day and have other roles for the other three cases. What is the probability that she will get her wish?

5. What is the probability that Mari will be a defense attorney exactly once during the four days? Explain your answer.

Baseball and Counting

You will now return briefly to the central problem and find the probabilities for a few more cases.

Then you will investigate some sophisticated ways to count. Among other things, you'll count the number of different ways to create ice cream cones. You'll also apply your new counting techniques to some more baseball outcomes.

Mike Holcombe is clearly confident that the list method has resulted in his finding all possible combinations.

And If You Don't Win 'em All?

The Good Guys would like to win all of their remaining seven games, but winning six out of seven wouldn't be so bad. The Bad Guys would also be pretty happy to win six out of seven.

1. Find the probability that the Good Guys will win the first six of their remaining games and then lose the seventh game.

2. Question 1 involves only one of several ways the Good Guys can compile a record of six wins and one loss in their remaining games. Find the probability that the Good Guys will win exactly six of their seven remaining games.

3. Find the probability that the Bad Guys will win exactly six of their seven remaining games.

But Don't Lose 'em All, Either

The activity *And If You Don't Win 'em All?* involves the "next-to-best" scenario for each of the two teams: winning six games and losing just one. Now you will examine the "next-to-worst" scenario.

1. Find the probability that the Good Guys will win the first of their remaining games and then lose the remaining six.

2. Find the probability that the Good Guys will win exactly one of their seven remaining games.

3. Find the probability that the Bad Guys will win exactly one of their seven remaining games.

The Good and the Bad

You have discovered a great deal about the probability of the Good Guys and the Bad Guys getting certain records.

For instance, you know the probability that the Good Guys will win six games and lose one, and you know the probability that the Bad Guys will win six games and lose one. But what is the probability that both things will happen? And for what other combinations can you find the probability?

You have already begun a chart of all possible combinations of individual records for the Good Guys and the Bad Guys. Use the information you have so far on the probabilities of some of these records to fill in as much of your chart as possible.

Top That Pizza!

Jonathan delivers pizza several evenings a week. He doesn't earn a lot, but he does get a free pizza for dinner every night he works.

The pizza shop has five toppings that Jonathan likes: pineapple, olives, mushrooms, onions, and anchovies.

1. Jonathan likes variety. If he always has exactly two of these five toppings on his pizzas, how many nights can he work without repeating a combination? Explain your answer.

2. Jonathan's sister Johanna also delivers pizza. She likes the same five toppings as her brother, but she always wants exactly *three* of them on her pizzas. How many nights can she work without repeating a combination? Explain your answer.

3. How are your answers to Questions 1 and 2 related? Explain why this relationship holds true.

Double Scoops

After Jonathan finishes his deliveries, he always treats himself to a two-scoop bowl of ice cream at the shop next to the pizza store. The ice cream shop serves 24 flavors of ice cream.

1. Jonathan always gets two different flavors for his two scoops of ice cream. How many different combinations of two scoops can he create?

2. Johanna always visits the ice cream shop after her deliveries, too. She likes her ice cream on a cone, and it's important to her which scoop is on top. After all, she says, eating chocolate and then vanilla is a different taste experience from eating vanilla and then chocolate.

Like her brother, Johanna always wants two different flavors. How many different two-scoop ice cream cones can she create?

Triple Scoops

Poor Johanna! She has had her tonsils out and is stuck at home with a sore throat. But fortunately, one of the things she can eat easily is ice cream.

Her friend Joshua is coming over, so she asks him to get her a three-scoop cone. She tells Joshua that he will probably find her brother Jonathan eating a bowl of ice cream at the shop. She asks Joshua to get her the same flavors that Jonathan has.

Sure enough, Jonathan is there—and he's actually having a three-scoop bowl of ice cream! He has one scoop each of pistachio, boysenberry, and chocolate, so Joshua knows to order those three flavors.

Unfortunately, neither Joshua nor Jonathan has any idea about the order in which Johanna will want her scoops. Joshua knows she is fussy about this, and he really wants her to get what she wants.

1. Joshua decides to get Johanna all possible cones with those three flavors. How many different cones does he buy?

2. Joshua realizes he is lucky that Johanna hadn't asked for a four-scoop cone. If she had requested a cone with pistachio, boysenberry, chocolate, and butter pecan without specifying the order of flavors, how many different cone possibilities would there be?

More Cones for Johanna

After Johanna recovers from her tonsillectomy, she continues to eat three-scoop ice cream cones. In fact, she thinks it would be fun to try every possible three-scoop cone the shop has to offer. Remember that the ice cream shop serves 24 flavors.

1. If Johanna has one three-scoop cone every day, how many days can she go before she will have to repeat?

2. Suppose instead that she has a different four-scoop cone each day. How long will it take for her to try them all?

3. Find a rule for determining the number of different cones in terms of the number of scoops on the cone. Base your work on the 24-flavor ice cream shop. Of course, your rule won't work for more than 24 scoops.

4. One day, while visiting relatives in another town, Johanna goes into a new ice cream shop. She figures out that there are 156 possible two-scoop cones that could be made from the flavors at this shop. How many different flavors does this shop serve?

Cones from Bowls, Bowls from Cones

In *Triple Scoops,* you saw that many different ice cream cones could be made from the scoops in a given bowl of ice cream. The number of cones depends on the number of scoops involved.

These questions continue that theme. (You do not need to figure out the number of flavors at each ice cream shop.)

1. At Francisco's Freeze, you can make 465 different two-scoop bowls of ice cream. How many different two-scoop ice cream cones can you make?

2. At Paige's Parlor, you can make 220 different three-scoop bowls of ice cream. How many different three-scoop ice cream cones can you make?

3. a. At Ashley's Ice Cream Shoppe, you can make 210 different four-scoop bowls of ice cream. How many different four-scoop ice cream cones can you make?

   b. At Carmen's Creamery, you can make 3024 different four-scoop ice cream cones. How many different four-scoop bowls of ice cream can you make? (*Careful:* This question and Question 4b reverse the situation presented in the other questions.)

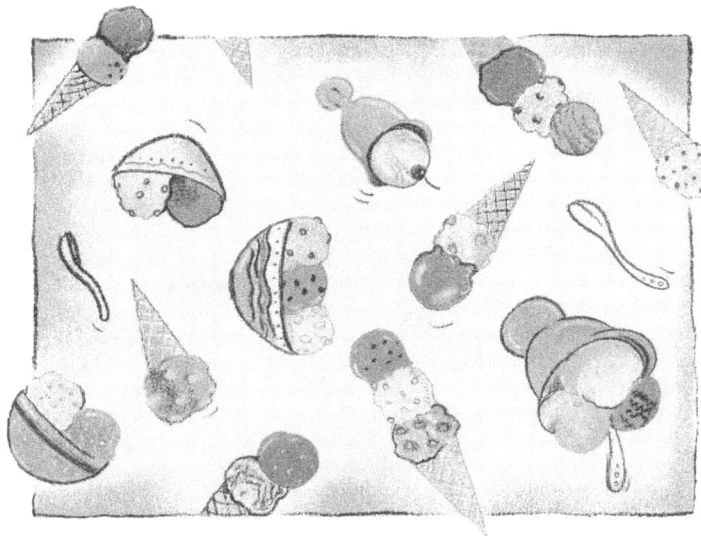

continued ▶

4. a. At Fiona's Flavors, you can make 792 different five-scoop bowls of ice cream. How many different five-scoop ice cream cones can you make?

   b. At Christopher's Cones, you can make 55,440 different five-scoop ice cream cones. How many different five-scoop bowls of ice cream can you make?

5. a. In general, if you know how many bowls of ice cream can be made with a given number of scoops, how do you find the number of different cones that can be made of that size?

   b. In general, if you know the number of cones that can be made with a given number of scoops, how do you find the number of different bowls that can be made?

Bowls for Jonathan

In *Cones from Bowls, Bowls from Cones,* you considered several ice cream shops. In some cases, you found the number of distinct bowls of ice cream of a particular size in terms of the number of distinct cones of that size. Apply the principles from that activity to these questions.

Johanna and Jonathan return to their favorite ice cream shop, which serves 24 flavors.

1. How many different three-scoop bowls of ice cream can they order?

2. How many different four-scoop bowls of ice cream can they order?

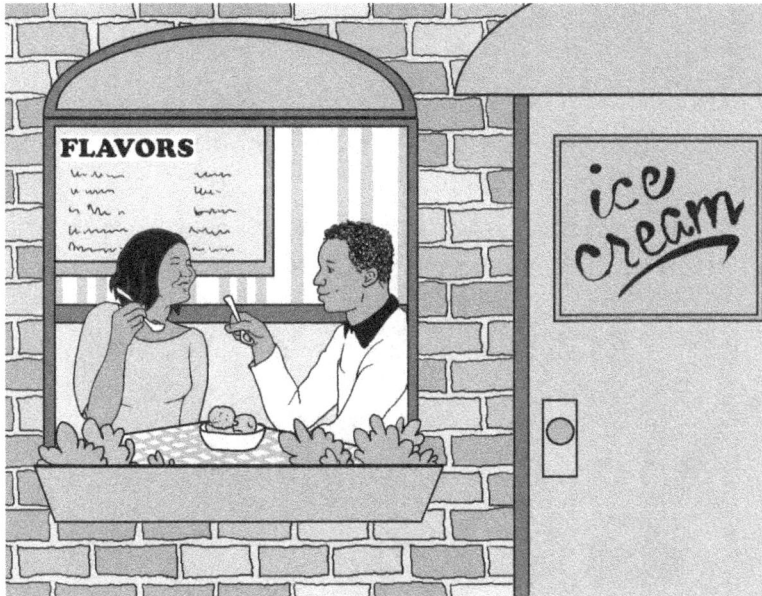

At the Olympics

1. The nation of Panacea will participate in the 400-meter race in the next Olympics. The nation has ten runners of equal ability. It must choose three of these runners to represent Panacea at the Olympics.

   How many different three-person teams are possible?

2. There will be ten finalists in the Olympics gymnastics competition. One of these ten will win the gold medal, one will win the silver, and one will win the bronze.

   After the competition, a plaque will be made listing the three winners in order. Use the fact that there are ten finalists to determine the number of possibilities for the sequence of names on this plaque.

3. Compare Questions 1 and 2.

   a. Which question is like an ice cream cone problem, and which is like an ice cream bowl problem?

   b. How can you find the answer to one of these two types of problems from the answer to the other?

Fair Spoons

○ The Original Problem

In the activity *Choosing for Chores*, Scott and Letitia determine who will wash the dishes and who will dry them by having Scott pull two spoons out of a bag.

The bag contains two spoons with purple handles and three with green ones. If the two spoons Scott pulls out are the same color, Scott will wash and Letitia will dry. If they are different colors, Letitia will wash and Scott will dry.

○ The New Problem

Letitia decides she doesn't like this method, because it turns out that she washes the dishes about 60% of the time. Scott thinks that if they find the right number of spoons of each color to put in the bag, they can make the probability of a match equal to 50%. But neither of them is sure what the right numbers would be.

What do you think? Find out as much about their choices as you can—don't merely find the simplest answer. Assume that plenty of spoons of both colors are available. As you work, keep track of the probability of a match in cases that do not come out to 50%.

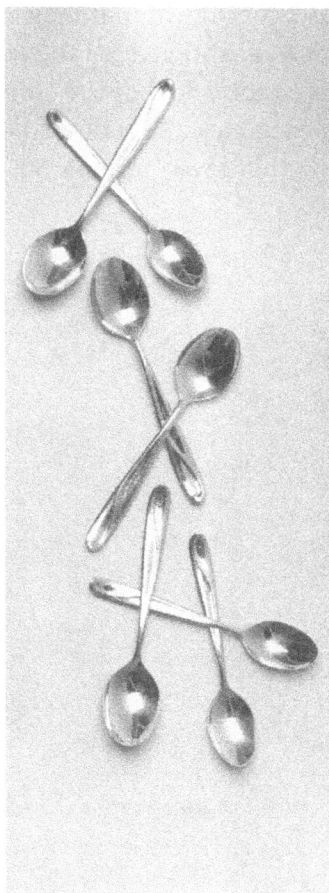

continued ◈

○ *Write-up*

1. *Problem Statement*

2. *Process*

3. *Solution:* Give the percentage of matches for all the specific examples you examined, and give all the combinations you found that led to matches exactly 50% of the time. Explain how you found the percentages in each case. Also describe any patterns you notice in the combinations that give matches 50% of the time.

4. *Self-assessment*

Which Is Which?

Permutations and Combinations

In recent activities, you have been carefully examining two types of counting problems. Mathematicians use the terms **permutations** and **combinations** to refer to these two types of problems.

We use the notation $_nP_r$ for permutation problems and the notation $_nC_r$ for combination problems. In both cases, the variable n stands for the size of the group you are choosing from. The variable r stands for the number of things you are picking from that group.

The numbers $_nC_r$ are called **combinatorial coefficients.** (There is no standard term for the numbers $_nP_r$.)

Your Task

In this activity, you will review the activities you have done so far in this unit.

1. Identify three specific problems you believe are combination problems and three specific problems you believe are permutation problems.

2. Explain why you think each problem is the type you say it is.

3. Express the answers to the questions in each problem you identified using the notations $_nP_r$ and $_nC_r$.

Formulas for $_nP_r$ and $_nC_r$

You have seen that many of the problems in this unit involve the concepts of combinations and permutations.

In *Which Is Which?*, you looked at how to apply these concepts and the notations $_nP_r$ and $_nC_r$ to problems in this unit. Now you will find formulas for $_nP_r$ and $_nC_r$.

1. Find formulas for these specific cases of $_nP_r$. Give each answer as an expression in terms of n. You may want to think about specific situations to which these cases apply.

   a. $_nP_1$

   b. $_nP_2$

   c. $_nP_3$

2. Find a general formula for $_nP_r$ in terms of n and r.

Next, think about the relationship between permutations and combinations in specific situations. For example, for a given number of ice cream flavors, what is the relationship between the number of three-scoop bowls and the number of three-scoop cones?

3. Develop a general equation expressing the relationship between $_nP_r$ and $_nC_r$.

4. Combine your results from Questions 2 and 3 to get a formula for $_nC_r$ in terms of n and r.

Who's on First?

The Good Guys have a terrific team. In fact, the team's manager, Sammy Lagrange, is sure they will win the pennant and go on to play in the World Series.

In the World Series, the first team to win four games is declared the world champion. (American baseball has always assumed that its best team is the best in the world.)

Sammy is already planning ahead, but one drawback of having such a good team is that it's often hard to make decisions.

1. When the Good Guys get to the World Series, Sammy will have to pick four different pitchers, one for each of the first four games of the World Series. He isn't thinking beyond four games, because he expects the Good Guys to win four games in a row and be declared the champions.

   Sammy has seven excellent pitchers to choose from. To prepare properly, the pitchers need to know which game, if any, they will pitch. This means Sammy has to determine not only which four pitchers to use but who will pitch in which game.

   He decides to put all possible sequences of four pitchers on slips of paper. He'll then pick one of these slips out of a baseball cap to settle the pitching order.

   How many slips of paper does Sammy need? (The answer is not seven! Sammy isn't doing this the easy way. Remember that each slip of paper shows a sequence with the names of four pitchers in order.)

continued ◗

And that's only the pitchers! For the rest of the team, Sammy is thinking only as far as the first game, but he still has choices to make.

2. To begin with, Sammy needs to choose his outfield—a left fielder, a center fielder, and a right fielder. He has five good outfielders to choose from, and they can each play any of these positions. Sammy needs to decide which outfielder will play which position.

   How many ways are there for Sammy to fill these positions for the first game?

3. Then there's the infield. Sammy needs a first baseman, a second baseman, a shortstop, and a third baseman. He has six highly qualified infielders. They are all versatile and able to play any of these four positions.

   How many possibilities must Sammy consider in filling the infield positions for the first game?

4. Fortunately, Sammy knows who his best catcher is. That player will complete his nine-player team for the first game. The nine players are a pitcher, three outfielders, four infielders, and the catcher.

   Considering all that's going on, how many different possibilities are there altogether for who will play each position for the Good Guys in the first game?

5. Once Sammy has decided on his nine players, he has to choose a batting order. That is, he needs to decide which of the nine players bats first, bats second, bats third, and so on.

   How many ways are there to arrange his starting nine players in the batting order for the first game?

Five for Seven

It's time to return to the central unit problem about the Good Guys and Bad Guys. There are still many possible outcomes with probabilities you have not yet figured out.

You will look at a few of those cases in this activity. You may find that you can apply what you learned about pizza and ice cream to the race for the baseball pennant.

1. What is the probability that the Good Guys will win five and lose two of their remaining seven games? Explain your answer.

2. What is the probability that the Bad Guys will win two and lose five of their remaining seven games? Explain your answer.

3. What is the probability that the outcomes from Questions 1 and 2 will both happen? That is, what is the probability that the Good Guys will win five games and lose two *and* the Bad Guys will win two games and lose five?

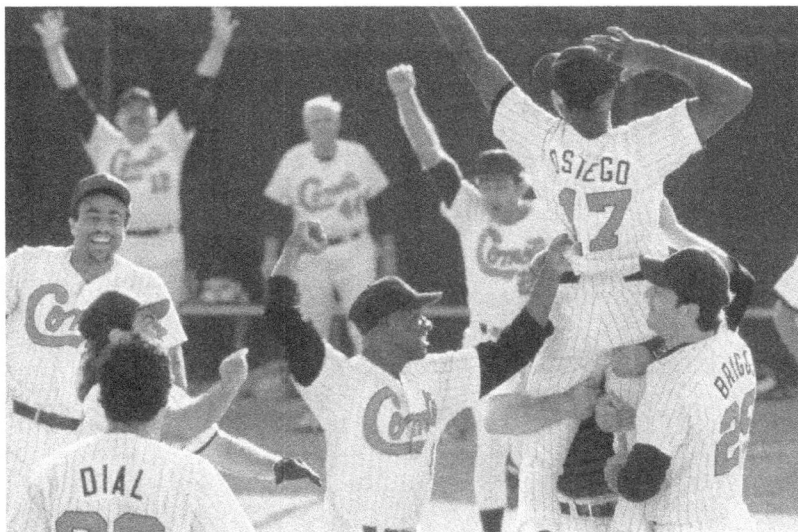

More Five for Sevens

In *Five for Seven,* you found two probabilities that are important in solving the unit problem.

- The probability that the Good Guys will win five and lose two of their remaining seven games
- The probability that the Bad Guys will win two and lose five of their remaining seven games

1. Find each of these similar probabilities.

   a. The probability that the Good Guys will win two and lose five of their remaining seven games

   b. The probability that the Bad Guys will win five and lose two of their remaining seven games

In *Five for Seven,* you also found the probability of both events happening—that is, the probability that the Good Guys will win five and lose two *and* the Bad Guys will win two and lose five. This probability represents only one of the 64 cells in your chart of possible records for the two teams.

2. Enter the values from Question 1 in your chart. Then find and record the probabilities for the three other cells that involve either two wins or five wins for each team.

Combinatorial Reasoning

Combinatorial coefficients can be helpful in many situations besides baseball and ice cream. Because probability involves counting cases, these special numbers play a role in a wide variety of probability problems.

In the next activities, you'll see how to use both permutations and combinations—along with principles of statistical reasoning, such as the null hypothesis—to find probabilities and to make decisions.

Annie Tam computes the number of ways the baseball teams can win or lose their remaining games.

What's for Dinner?

Lai Yee wants to buy a motor scooter. His parents encourage him to take care of his own expenses, so they offer to pay him for providing the family's dinner four nights a week: Sunday, Tuesday, Thursday, and Saturday. They will pay him $20 a week for his work, in addition to reimbursing him for the cost of the food.

Lai Yee has never cooked before, and his parents want to make sure they won't get the same one or two meals over and over again. So they decide he must submit his planned menu for each week in advance, showing the four meals for that week. His menus must meet two rules.

- No weekly menu can contain two of the same meal.

- Each weekly menu must be different from all previous weekly menus.

Lai Yee is clever. After some research, he discovers that there are seven meals already prepared that he can buy from nearby restaurants. That way, he won't have to learn to cook. The seven meals are Chinese noodles, pasta, tacos, quiche, pizza, sushi, and roasted chicken.

1. How much money can Lai Yee earn before he has to learn to cook something? Express your answer using the notation $_nP_r$ or $_nC_r$ appropriately. Also find the actual numeric answer, and explain how you got your answer.

continued ▶

2. There was a misunderstanding between Lai Yee and his parents. He thought he could use the same set of four meals for more than one weekly menu if he simply presented them in a different order. To his parents, "different weekly menu" means "a different set of four meals."

   Whichever way you interpreted Question 1, now answer the question using the other interpretation. Again, express your answer using the notation $_nP_r$ or $_nC_r$ appropriately, and also find the actual answer.

3. What if Lai Yee were to cut back to three meals per week? How many weeks could he go under his interpretation before he has to cook something? Under his parents' interpretation?

All or Nothing

In most cases, the formulas for the combinatorial and permutation coefficients are pretty clear. In some special cases, though, it helps to have a situation to give concrete meaning to these numbers.

1. You found that there are 21 different sequences by which the Good Guys can win exactly five of their remaining seven games. In other words, the combinatorial coefficient $\binom{7}{5}$ is equal to 21. Remember that the notation $\binom{n}{r}$ means the same thing as $_nC_r$.

   Use the pennant race situation to determine the value of the combinatorial coefficient $\binom{7}{0}$. Explain your reasoning.

2. You found that there are ten different pizza combinations Johanna can create if she chooses three toppings for her pizza out of the five she likes. In other words, the combinatorial coefficient $\binom{5}{3}$ is equal to 10.

   Use the pizza situation to determine the value of the combinatorial coefficient $\binom{5}{5}$. Explain your reasoning.

3. Use either Jonathan's or Johanna's ice cream preferences to determine the values of $_{24}P_0$ and $\binom{24}{0}$. Explain your reasoning.

4. Use any of the situations in Questions 1 to 3, or another situation you make up, to determine the values of $_nP_1$ and $\binom{n}{1}$. Explain your reasoning.

The Perfect Group

At last, in his third year of high school, Julio has lucked out. He is finally part of what he considers the perfect four-person study group in his class. This is the group he would have chosen for himself on the first day of his first year of high school if he'd had a chance.

Each time new groups were formed, he hoped for this group. Sometimes he'd get one or two members of this ideal group, but never all three.

How lucky was Julio? Begin with these assumptions.

- There are 32 students in Julio's class.
- This 32-person class has been together throughout high school so far.
- New groups are created randomly every two weeks.
- Groups always have four students.

1. Guess the probability of Julio getting his perfect group sometime in three years. Resist the temptation to do any arithmetic yet—simply give your intuitive idea about what the chances are.

2. Now actually find this probability. You will need to make some further assumptions to do so. State those assumptions clearly.

And a Fortune, Too!

The King's New Scale

Do you remember the economical king from the Year 1 POWs *Eight Bags of Gold* and *Twelve Bags of Gold* (in the unit *The Pit and the Pendulum*)? Well, he's still around, and despite his economical nature, he has only five bags of gold now.

One reason he has less gold is that his old pan-balance scale broke down. His adviser found an antique scale to replace it—the kind in which you drop in a penny and are told your weight. This scale doesn't let the king compare weights, but it does give very precise measurements.

The king was very excited about another feature of the scale. With each weight, he also got a slip of paper that predicted his fortune. But he didn't pay attention to the fact that he was spending a penny every time he used the scale. He spent a lot of pennies before he realized his fortune had dwindled to five bags of gold.

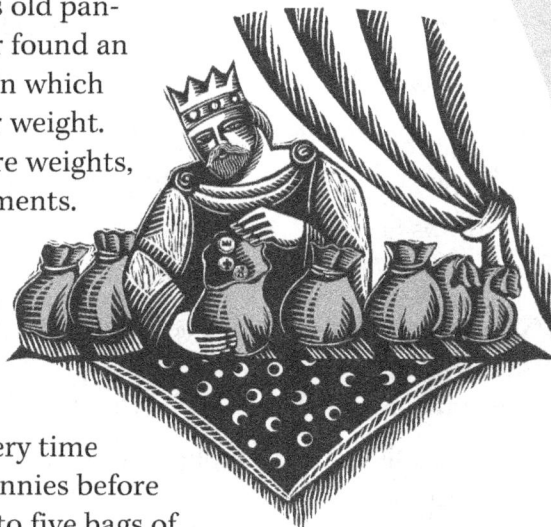

The King's New Problem

Now, as before, the king has given one bag of gold to each of the five people he trusts the most in his kingdom. And, as before, rumors have drifted back to the king that one of these five caretakers is not to be trusted. According to rumor, this person is asking a counterfeiter to make phony gold.

The king has brought in the local counterfeiter for questioning. He wants to find out who hired her to double-cross him.

continued

She admits being involved, but refuses to name the traitor. She won't even say whether the counterfeit gold she is making is heavier or lighter than real gold. All the king learns from her is that one of the five bags is now filled with counterfeit gold and that this bag weighs a different amount from the others.

○ Your Challenge

The king wants to use his scale to determine two things.

- Which bag weighs a different amount
- Exactly how much that bag weighs

Of course, he wants to learn these things economically, using the fewest pennies possible. (He has suddenly started to economize—even with pennies!)

The court mathematician says it can be done by using only three pennies. No one else sees how it can be done with so few weighings. Anyone can do it with five pennies. Some say they can do it with four pennies. But three pennies? Can you figure it out?

○ Write-up

1. *Problem Statement*

2. *Process*

3. *Solution:* You might get only a partial solution to this problem. If so, explain what cases your solution covers and where you got stuck.

4. *Self-assessment*

Feasible Combinations

In the unit *Meadows or Malls?*, you generalized ideas about graphs and inequalities to solve linear programming problems in several variables.

One important principle from that unit is that you can find the corner points of a feasible region by examining certain systems of linear equations. If the set of constraints uses n variables, then each system you examine should contain n linear equations.

In this activity, you'll apply ideas from this unit to find how many linear systems you might have to consider to solve linear programming problems.

1. The central problem from *Meadows or Malls?* involves six variables. They are labeled G_R, A_R, M_R, G_D, A_D, and M_D. The situation is described by these 12 constraints.

I	$G_R + G_D = 300$	VII	$G_R \geq 0$
II	$A_R + A_D = 100$	VIII	$A_R \geq 0$
III	$M_R + M_D = 150$	IX	$M_R \geq 0$
IV	$G_D + A_D + M_D \geq 300$	X	$G_D \geq 0$
V	$A_R + M_R \leq 200$	XI	$A_D \geq 0$
VI	$A_R + G_D = 100$	XII	$M_D \geq 0$

Each inequality in this list has a corresponding linear equation. For instance, the inequality $G_D + A_D + M_D \geq 300$ corresponds to the linear equation $G_D + A_D + M_D = 300$.

So the 12 constraints lead to 12 linear equations. Because the *Meadows or Malls?* problem involves six variables, every corner point for the feasible region is the solution to a system that consists of six of these 12 equations.

a. How many six-equation systems can you form from the 12 equations? Of course, some of these systems do not actually lead to a corner point of the feasible region.

continued

If any of the constraints in a linear programming problem are actually equations, then all corner points have to fit those equations. Therefore, in your search for corner points, you can restrict yourself to linear systems that include those equations.

b. In the *Meadows or Malls?* problem, four of the constraints are equations. Suppose you examine only six-equation systems that include these four equations. That is, you examine only systems consisting of the four constraint equations together with two of the eight equations that correspond to constraint inequalities.

How many systems will you need to consider?

2. Suppose a linear programming problem has eight variables and 20 constraints, and that three of the constraints are equations.

How many linear systems would you need to consider? Again, some of these systems might not actually lead to a corner point of the feasible region.

About Bias

At Bayside High, a 15-member committee handles many decisions. The school-wide committee consists of ten adults and five students. Principal Fifer has been asked to select a special subcommittee of six people out of this group of 15.

Students are furious because they just learned that the subcommittee consists entirely of adults. They feel that the principal stacked the subcommittee with adults and didn't consider students. Principal Fifer, however, claims the subcommittee was chosen randomly.

The students have decided to present their case to the school board.

1. As part of their presentation, they want to tell the school board the probability of getting only adults if the principal had selected six people at random from the committee of 15. Find this probability, giving your answer both as a number and as an expression using the notation $_nP_r$ or $_nC_r$ appropriately.

2. Do you think the principal stacked the committee? Explain your answer.

Binomial Powers

You've seen that combinatorial coefficients can be helpful in finding probabilities like those involved in the unit problem. As you know, these numbers are also called **binomial coefficients.**

(*Reminder:* A *binomial* is an expression that is the sum of two terms, each of which is a product of numbers and variables. For example, $3x + 2y$, $5 - z$, $-3xy + 15zw$, and $7x^3 + 3xy^2$ are binomials.)

A little later in this unit, you'll find out what binomials have to do with the combinatorial coefficients $_nC_r$. In preparation for that discussion, your task now is to simplify certain powers of binomials by writing each expression as a sum of terms, without parentheses.

The diagram illustrates Question 1. You may find similar diagrams helpful for Questions 2 through 6 (and perhaps even for Question 7).

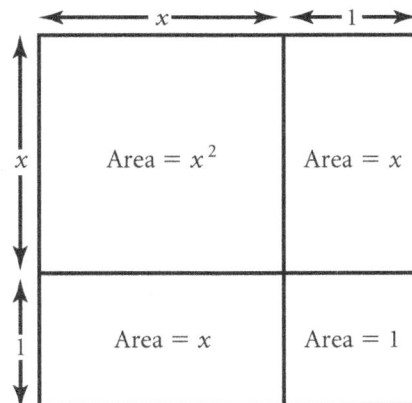

1. $(x + 1)^2$

2. $(a + b)^2$

3. $(r - 7)^2$

4. $(5g + 1)^2$

5. $(K + 4L)^2$

6. $(3y - 8)^2$

7. $(a + b)^3$

As long as you're simplifying algebraic expressions, go ahead and simplify these (which are not powers of binomials).

8. $x^2(2x^3 + 3x - 6)$

9. $(x^2 + 3)(x^2 - 2x + 4)$

Complete the square for each of these expressions. That is, find a value for c so that the expression is the square of a binomial.

10. $x^2 + 8x + c$

11. $x^2 - 13x + c$

Don't Stand for It

In a very small town is a small factory with ten workers. All ten workers go out for lunch every day, and they all go out at the same time. No one else in town goes out for lunch.

There are two restaurants in town. Every day, each worker randomly chooses to eat at one restaurant or the other. Both establishments have only counter seating (no tables).

One owner has ten stools at her counter just in case all ten workers visit her restaurant on the same day.

The other owner realizes it is unlikely all ten will come on the same day. He figures it's okay for business as long as he isn't short of stools more than an average of once a month. Because there are typically 20 workdays per month, he wants enough stools so that the chance of running short on any given day is less than 5%.

How many stools should he provide? Justify your answer.

Adapted from *Introduction to Finite Mathematics* by John G. Kemeny, J. Laurie Snell, and Gerald L. Thompson (Englewood Cliffs, NJ: Prentice-Hall, 1957).

Stop! Don't Walk!

Patience Walker always walks to school using the same route, although the time of day varies. She thinks one of the stoplights on the route to school has it in for her. It always seems to be red when she approaches the corner, no matter what time of day it is. She thinks it happens too often to be a coincidence.

Patience is actually not very patient, so she is anxious to get to the bottom of this.

She phones the Department of Public Works and is told that within the traffic light's timing cycle, the light is set to be red 60% of the time. Patience finds this difficult to believe and asks the DPW to investigate the light. The DPW representative tells her they have little time for such trivial matters and asks her to call back when she has some hard evidence.

1. Patience keeps track of the light for five days (one school week). Sure enough, the light is red on her way to school every one of those days.

   If the information from the DPW is correct, what is the probability of that happening? Explain your answer. Assume that Patience is equally likely to arrive at the light at any point during its cycle.

2. Patience is afraid the DPW won't be convinced by a five-day survey, so she keeps track for two more school weeks, for a total of 15 days. She finds the light to be red on 13 of those 15 days.

   She's ready to confront the DPW. If the light is really red exactly 60% of the time during each cycle, what is the probability that Patience would find it red 13 or more times out of 15?

Pascal's Triangle

Pascal's triangle is an array of numbers that contains many interesting patterns and relationships. The array is named for a French mathematician who did important work in the theory of probability, but this arrangement of numbers was studied long before his time.

In the upcoming activities, you'll investigate how the array is formed and see some of its many applications.

In creating Pascal's triangle, Maile Martin, Jeanette Austria, and Kahala Neil find many patterns.

Pascal's Triangle

The triangular arrangement shown here is the beginning of a pattern of numbers commonly called **Pascal's triangle.** Although only six rows are shown, the pattern can be extended indefinitely.

```
                    1
                 1     1
              1     2     1
           1     3     3     1
        1     4     6     4     1
     1     5    10    10     5     1
   ?     ?     ?     ?     ?     ?     ?
 ?     ?     ?     ?     ?     ?     ?     ?
```

This number pattern is named in honor of the French mathematician Blaise Pascal (1623–1662), who developed the beginnings of the modern theory of probability.

Though Pascal was a distinguished mathematician, he was also famous as a physicist, geometer, and religious philosopher. Among other things, he invented the first digital calculator. A well-known computer language is named for him as well.

Pascal was not the first person to work with this numeric pattern. In fact, the pattern has been found in use as early as around 1300 CE, in a book of Chinese prints.

1. Find a pattern in Pascal's triangle that will allow you to extend it to more rows. Then use this pattern to extend Pascal's triangle to at least ten rows altogether. (Save this extended version of Pascal's triangle, because you will need it for the rest of the unit.)

2. Find other patterns in the triangle. Describe each new pattern you find in words and with examples.

Hi There!

The handshake problem is a classic mathematics problem. It goes like this.

There are n people in a room. Everyone shakes hands exactly once with everyone else. How many handshakes are there?

When two people shake hands, it counts as one handshake.

1. Find the answer to the handshake problem for the case $n = 3$. Also find the answers for $n = 5$, for $n = 10$, and for two other specific cases of your choice.

2. Look for a pattern in your answers, or find a general formula for n people.

3. Explain how this problem appears to be related to Pascal's triangle.

4. How is this problem related to combinatorial coefficients? Remember that the combinatorial coefficient $_nC_r$ tells how many different bowls of ice cream you can make with r scoops (of different flavors) if there are n flavors altogether.

Pascal and the Coefficients

The entries in Pascal's triangle, it turns out, are combinatorial coefficients. In fact, this is the main reason Pascal's triangle is important in mathematics.

It is standard practice to refer to the top row of Pascal's triangle as "row 0," the next row as "row 1," and so on. Similarly, the first number in each row is called "entry 0," the next number is "entry 1," and so on. This numbering system connects the position of a number to its meaning as a combinatorial coefficient.

For example, according to this system, the boxed number 10 shown here is entry 2 of row 5 of Pascal's triangle. This relates to the fact that 10 is equal to the combinatorial coefficient $\binom{5}{2}$, which tells how many different bowls of ice cream you can make with two scoops (of different flavors) if there are five flavors altogether.

```
                    1
                  1   1
                1   2   1
              1   3   3   1
            1   4   6   4   1
          1   5  [10] 10   5   1
        ?   ?   ?   ?   ?   ?   ?
      ?   ?   ?   ?   ?   ?   ?   ?
```

In general, it can be proved that entry r of row n is the combinatorial coefficient $\binom{n}{r}$. For instance, the row 1 4 6 4 1 is row 4 and consists of the combinatorial coefficients $\binom{4}{0}$, $\binom{4}{1}$, $\binom{4}{2}$, $\binom{4}{3}$, and $\binom{4}{4}$.

continued

1. Check that $\binom{4}{0}$, $\binom{4}{1}$, $\binom{4}{2}$, $\binom{4}{3}$, and $\binom{4}{4}$ do have the values 1, 4, 6, 4, and 1, respectively. Explain the values in terms of bowls of ice cream.

2. Use the connection between Pascal's triangle and combinatorial coefficients to find these values.

   a. $\binom{6}{5}$

   b. $\binom{7}{4}$

   c. $\binom{9}{5}$

   d. $\binom{10}{6}$

3. One feature of Pascal's triangle is that each row begins and ends with the number 1. In terms of combinatorial coefficients, this means $\binom{n}{0}$ and $\binom{n}{n}$ are both equal to 1, for any value of n.

   Explain this feature of Pascal's triangle in terms of bowls of ice cream or using some other model for combinatorial coefficients.

Combinations, Pascal's Way

In the activity *Pascal's Triangle,* you explored patterns and relationships in that special array of numbers. You now know that the numbers in Pascal's triangle are actually combinatorial coefficients.

Your task now is to examine the patterns and relationships in Pascal's triangle in light of the connection between Pascal's triangle and combinatorial coefficients.

For each pattern or relationship you found in Pascal's triangle, do two things.

- Express the pattern or relationship in terms of combinatorial coefficients.

- Explain the pattern or relationship based on the meaning of combinatorial coefficients—for example, by using bowls of ice cream or another model for $_nC_r$.

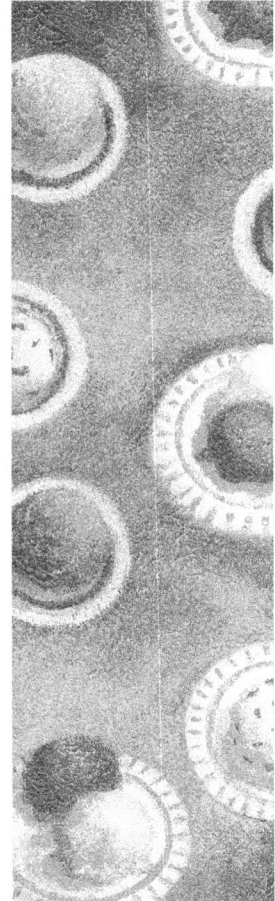

Binomials and Pascal—Part I

In *Binomial Powers,* you found powers of some binomials. Now you will look at powers of a special binomial and examine how the coefficients in the expansion are related to Pascal's triangle.

1. Expand and simplify each of these expressions, combining terms and writing each result as a sum without parentheses.

   a. $(a + b)^2$

   b. $(a + b)^3$

   c. $(a + b)^4$

   d. $(a + b)^5$

2. Examine the coefficients in your results. How are they related to Pascal's triangle?

Blaise Pascal (1623–1662)

Binomials and Pascal—Part II

In *Binomials and Pascal—Part I*, you examined powers of the binomial $a + b$, expanding expressions of the form $(a + b)^n$.

In each case, the coefficients form a row of Pascal's triangle. That is, they are numbers of the form $\binom{n}{r}$.

Now you will use the connection between the coefficients and Pascal's triangle to find the expansions for powers of some other binomials.

1. Find the expansion of $(a + b)^{10}$. Using your insights from the previous activity and your copy of Pascal's triangle, write the expansion as a sum of terms with powers of a and b and with appropriate coefficients from Pascal's triangle.

2. Find the expansion of $(a + 2)^5$.

3. Find the expansion of $(x - 1)^4$. Be careful about signs. It might help to think of $x - 1$ as $x + (-1)$.

A Pascal Portfolio

Write a summary of what you know about Pascal's triangle. Include these elements.

- How to create Pascal's triangle
- How to use Pascal's triangle to find combinatorial coefficients
- Properties of Pascal's triangle and explanations of these properties in terms of combinatorial coefficients
- Why the numbers in Pascal's triangle are called *binomial coefficients*

Another portrait of Blaise Pascal

The Baseball Finale

After many digressions, it's now time to solve the central problem of the unit. Remember—in this unit as in baseball— it's not over 'til it's over!

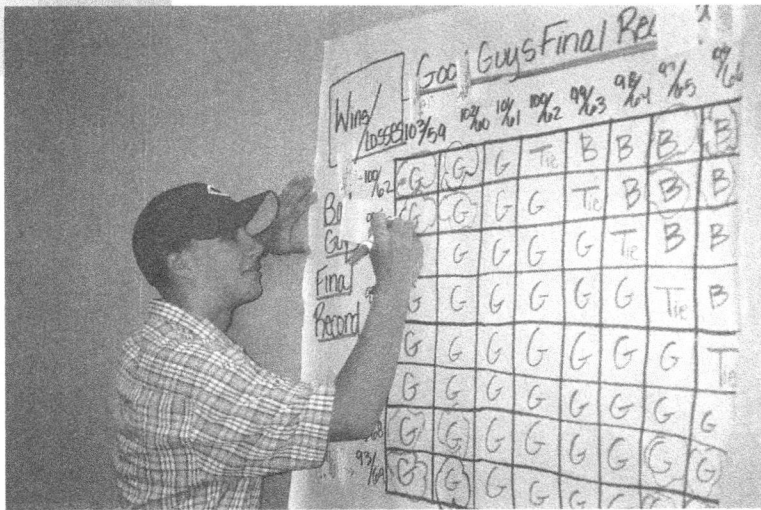

Jeff Kirilov records the probability of one of several possible outcomes for the teams competing in the pennant race.

Race for the Pennant! Revisited

Time is running out, and the season will soon be over. Here's your last chance to figure out the probability that the Good Guys will win the pennant.

Once again, here are the records of the two teams.

Team	Games won	Games lost	Games left
Good Guys	96	59	7
Bad Guys	93	62	7

Recall these key facts and assumptions.

- The Good Guys and Bad Guys do not play each other.
- In each game the Good Guys play, their probability of winning is .62.
- In each game the Bad Guys play, their probability of winning is .6.

Determine the probability that the Good Guys will win the pennant. Also determine the probability that the teams will be tied when they each finish the remaining seven games.

Graphing the Games

You've determined that there are eight possible overall outcomes for the Good Guys' final seven games. They can win all seven games, they can win six and lose one, and so on, down to losing all seven. You've also seen that these eight outcomes are not equally likely.

1. Make a bar graph showing the probabilities for the eight possible overall outcomes. Keep in mind that in each game they play, the Good Guys' probability of winning is .62.

The Bad Guys are almost as good a team as the Good Guys. For each remaining game, their probability of winning is .6. If you were to make a bar graph showing the probability of each overall outcome, it would not be much different from the graph for the Good Guys.

A fair coin, on the other hand, has a probability of .5 of coming up heads. Suppose you flipped a fair coin seven times and recorded the results. As with the baseball problem, there are eight possible overall outcomes (seven heads and no tails, six heads and one tail, and so on, down to no heads and seven tails).

2. Make a bar graph showing the probabilities for the eight possible overall outcomes for seven flips of a fair coin.

3. Discuss the similarities and differences between your two graphs.

Binomial Probabilities

In the central unit problem, the Good Guys play a series of seven games. For each game, the result is either a win or a loss, and the team's probability of winning is .62.

In a sequence of seven flips of a fair coin, the result of each flip is either heads or tails, and the probability of heads is .5 for each flip.

Although the probabilities are different in the two examples, the situations share two key features.

- There is some "event" with two possible outcomes.
- The event is repeated some number of times, but the probabilities of the two outcomes do not change. That is, the result of one repetition does not influence the results of other repetitions, so each occurrence of the event is *independent* of previous occurrences.

The probabilities associated with such a situation define the **binomial distribution.** Just as there are many variations of the normal distribution, with different means and standard deviations, there are also many variations of the binomial distribution. These variations depend on the probabilities for the two outcomes and the number of times the event is repeated.

1. The Good Guys are thinking about next year. Suppose their outcome for next season follows a binomial distribution. Specifically, assume that as in the unit problem, they have a probability of .62 of winning each time they play a game.

   If their season consists of 162 games, what is the probability that they will win exactly 100 games? Express your answer using a combinatorial coefficient. You do not need to get a numeric value for the probability.

continued ▶

2. Now consider the general case of the binomial distribution. As with the baseball and coin-flip examples, the event in question must have two outcomes. The two outcomes are generically referred to as *success* and *failure,* and each repetition of the event is called a *trial.* For instance, in the central unit problem, the event is a single game, success means winning the game, failure means losing the game, and the part of the Good Guys' season under consideration involves seven trials.

Suppose the probability of success is p and the event is repeated n times.

a. What is the probability of failure on each trial?

b. How many failures will there be if there are r successes?

c. What is the probability of getting exactly r successes out of the n trials?

Pennant Fever Portfolio

You will now put together your portfolio for *Pennant Fever*. This process has three steps.

- Write a cover letter that summarizes the unit.
- Choose papers to include from your work in the unit.
- Discuss your personal growth during the unit.

Cover Letter

Look back over *Pennant Fever* and describe the central problem of the unit and the key mathematical ideas. Your description should give an overview of how key ideas like combinations, permutations, and Pascal's triangle were developed and how they were used to solve the central problem. Also include ideas that were not directly part of the unit problem, such as the birthday problem and the binomial distribution.

In compiling your portfolio, you will select some activities you think were important in developing the unit's key ideas. Your cover letter should include an explanation of why you selected each item.

continued ▸

Selecting Papers

Your portfolio for *Pennant Fever* should contain
these items.

- *Diagrams, Baseball, and Losing 'em All*
- *Which Is Which?*

  Include in your portfolio the activities you discussed
  as part of this activity.

- *A Pascal Portfolio*
- *"Race for the Pennant!" Revisited*
- *Binomial Probabilities*
- An activity illustrating the use or meaning of the
  binomial theorem
- A Problem of the Week

  Select one of the POWs you completed in this unit (*Happy Birthday!*,
  Let's Make a Deal, Fair Spoons, or *And a Fortune, Too!*).

- Other key activities

  Identify two concepts you think were important in this unit. For each
  concept, choose one or two activities that helped your understanding
  improve, and explain how the activity helped.

Personal Growth

Your cover letter for *Pennant Fever* should describe how the
mathematical ideas were developed in the unit. In addition, write
about your personal development during this unit. Also include any
thoughts about your experiences that you wish to share with a reader
of your portfolio.

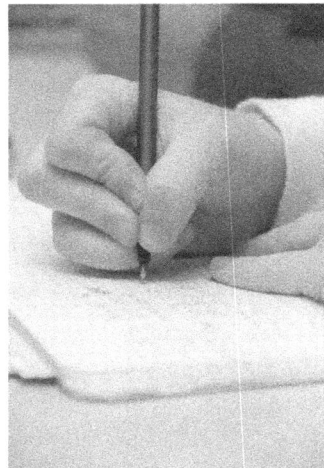

Supplemental Activities

Probability and counting techniques form the heart of this unit, and many of the supplemental activities continue those themes. Others follow up on ideas from the POWs. Here are some examples.

- *Programming a Deal* and *Simulation Evaluation* build on the activity *Simulate a Deal*.
- *Determining Dunkalot's Druthers* is a basketball probability problem quite similar to the central unit problem.
- *Sleeping In* and *My Dog's Smarter than Yours* are probability problems involving the combinatorial coefficients.

Putting Things Together

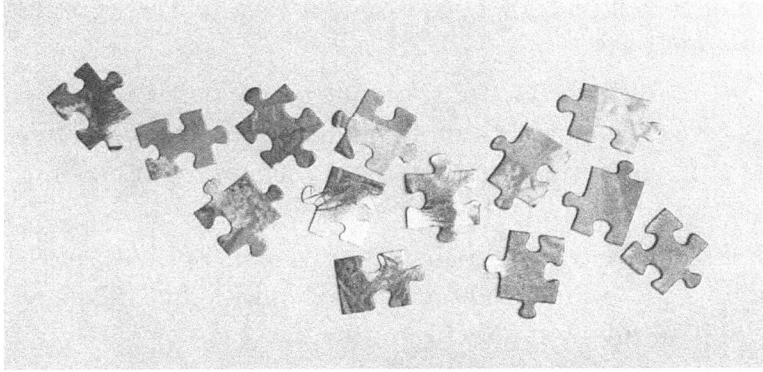

In the activity *Possible Outcomes,* you saw that the Good Guys can end up with any of eight possible records for their final seven games—from seven wins and no losses to no wins and seven losses. Similarly, there are eight different possibilities for the Bad Guys' final seven games. You also used this information to find the number of possible combinations of records for the two teams.

Here are some other problems involving finding the number of possibilities by making combinations from different lists.

1. The school cafeteria offers three main dishes (tacos, pasta, and chef's salad), four beverages (milk, iced tea, orange juice, and apple juice), and two desserts (peaches and pineapple). If you choose one main dish, one beverage, and one dessert, how many options do you have for creating your meal?

2. You're planning your weekend. You've decided to go to a movie Friday night, a concert Saturday night, and a sports event Sunday afternoon. There are five good movies showing, four excellent bands performing, and six local teams playing. How many different entertainment plans are available to you?

continued ◗

3. You will be away for 60 days and are planning what clothes to bring with you. For each day, you will need a shirt, a pair of pants, a pair of shoes, and a hat. You don't mind wearing an individual item more than once, but you don't want to wear the exact same outfit on two different days.

   You realize that if you bring one shirt, one pair of pants, one pair of shoes, and 60 hats, you could simply change hats each day to create exactly 60 different outfits, but that doesn't seem very interesting. Besides, all those hats would be pretty bulky!

   a. How many different shirts, pairs of pants, pairs of shoes, and hats should you bring? Find several possibilities that will give you exactly 60 different possible outfits.

   b. Find a combination that will require as few total items as possible.

4. What do Questions 1 to 3 have in common? Formulate some general principles for dealing with situations like this.

Ring the Bells!

After your experience with *Go for the Gold!,* you've decided that television game shows could be lots of fun. You apply and are accepted to be a contestant on *Ring the Bells!* You find out that you must pay a $200 fee to be on the show.

The show involves two buttons that are connected to colored bells and buzzers through a computer that uses a random number generator. Here's how the game works.

You begin by pushing button 1. There is a 40% chance that a green bell will ring, a 20% chance that a yellow bell will ring, and a 40% chance that a red buzzer will ring. If the green bell rings, you go on to button 2. If the red buzzer rings, the game is over and you lose and get no prize. If the yellow bell rings, you get a second try with button 1.

If you push button 1 a second time, the yellow changes from a bell to a buzzer, but the probabilities stay the same. If you get red or yellow (the buzzers), the game is over and you get a consolation prize of $100. If you get the green bell, you go on to button 2.

With button 2, there are only two possible outcomes: the green bell and the red buzzer. You have a 40% chance of getting the green bell and a 60% chance of getting the red buzzer. If you get the bell, you win $1,000. If you get the buzzer, the game is over and you lose and get no prize.

Do you accept the invitation to be on the show? Use both an area model (or sequence of area models) and a tree diagram to explain your decision. And don't forget about the $200 fee!

Programming a Deal

In the activity *Simulate a Deal,* you tested both the "switch" and "stay" strategies by carrying out a simulation. Like many simulations, this one can be accomplished by a calculator or computer program using a random number generator.

Your task now is to write such a program. Here are two options you might consider.

- **One-game-at-a-time program:** In this type of program, the user decides at the start of each game which of the two strategies to test. The program uses a random number generator to decide where the car is and then allows the user to choose the initial door. The game might end with a message like, "You decided to <'switch' or 'stay'> and you ended up with <'a car' or 'a worthless prize'>." With this type of program, the user has to keep track of the results.

- **Many-games-at-a-time program:** In this type of program, the user decides which strategy to test and also states how many games to play. The program then plays that many games one after another, using a random number generator to decide both where the car is and what the player's initial guess is. At the end of the game, the program gives a message like, "You used the <'switch' or 'stay'> strategy <some number> times. You got the car <some number> times and a worthless prize <some number> times."

To write your program, use one of these options or come up with a variation of your own.

Simulation Evaluation

When your class compiled its results from *Simulate a Deal*, you probably concluded that the "switch" strategy was better than the "stay" strategy. But whenever you use a simulation to estimate probabilities, it's a good idea to ask yourself how reliable your estimates are.

If you had no evidence about the strategies, you might take as your null hypothesis that the two strategies are equally good. Under this null hypothesis, you would expect the number of successes for the "switch" strategy to be the same as the number of successes for the "stay" strategy. (That's *not* the same as saying that each strategy has a 50% chance of success.)

Your class data probably did not show the two strategies having the same rate of success. Your task now is to evaluate whether the difference between your class results and the results expected under the null hypothesis just described might be due to sampling fluctuation.

Specifically, answer this question.

If the two strategies are actually equally good, what is the probability of getting results as far off from equal for the two strategies as your actual class results?

Use the chi-square statistic to compare two populations: games played using the "switch" strategy and games played using the "stay" strategy. Treat the overall class results from simulations for each strategy as your sample from that population.

The Chances of Doubles

When you roll a pair of dice, getting a *double* means getting the same result on both dice.

1. Suppose you roll a pair of ordinary dice. Explain why the probability of rolling a double is $\frac{1}{6}$.

Now suppose you roll the pair of dice and then roll the pair again. You have a $\frac{1}{6}$ chance of getting a double the first time. You also have a $\frac{1}{6}$ chance of getting a double the second time. So it might seem reasonable that the probability of getting a double on at least one of the rolls would be $\frac{1}{6} + \frac{1}{6}$.

By that reasoning, if you rolled the pair of dice three times, your probability of getting a double on at least one of the rolls would be $\frac{1}{6} + \frac{1}{6} + \frac{1}{6}$. And if you rolled the pair six times, the probability would be $\frac{1}{6} + \frac{1}{6} + \frac{1}{6} + \frac{1}{6} + \frac{1}{6} + \frac{1}{6}$, which equals 1.

In other words, by this reasoning, you would be certain of getting a double on at least one of the six rolls—which is definitely not true.

2. a. Explain why the probability of getting a double on at least one of two rolls of the pair of dice is not simply $\frac{1}{6} + \frac{1}{6}$.

   b. Find the correct probability of getting a double on at least one of two rolls of the pair of dice.

3. What is the probability of rolling a double at least once if you roll the pair of dice three times? If you roll four times? If you roll *n* times?

Determining Dunkalot's Druthers

Tyler Dunkalot's team has tied with another team for the basketball league's championship. He and the other team's captain have to choose between two options for determining the champion.

- **Option 1:** A three-game series between the two teams, in which the first team to win two games is the champion

- **Option 2:** A five-game series between the two teams, in which the first team to win three games is the champion

Tyler estimates that in each game the teams play, his team has a probability of .55 of winning. His team has been getting better over the season, and the other team has stayed pretty much the same.

Which of the two methods for choosing the champion—two out of three or three out of five—would you advise Tyler to support?

Sleeping In

Cynthia's school has eight class periods, which follow this schedule.

1st period:	8:00–8:50
2nd period:	9:00–9:50
3rd period:	10:00–10:50
4th period:	11:00–11:50
5th period:	12:00–12:50
6th period:	1:00–1:50
7th period:	2:00–2:50
8th period:	3:00–3:50

Each student takes five courses and has three free periods. The free periods are assigned randomly, and students do not need to report to school until the period when their first course meets

Cynthia does not think she should be expected to be coherent at 8:00 a.m. Even 9:00 a.m. is a bit early for her to think clearly. She wonders what her chances are of being able to sleep late.

1. What is the probability that Cynthia will have the first period free?

2. What is the probability that Cynthia will have the first two periods free?

3. What is the probability that Cynthia will get her dream schedule and have the first three periods free?

Twelve Bags of Gold Revisited

If you did the *Twelve Bags of Gold* POW in Year 1, perhaps you can now improve on your previous work. Maybe you can give a clearer explanation or a simpler solution. Here's the original problem.

The king has 12 bags of gold. Each of his 12 bags holds exactly the same amount of gold as each of the others, and they all weigh the same. Except . . .

Rumor has it that one of his 12 trusted caretakers is not so trustworthy. Someone is making counterfeit gold. So the king sends his assistants to find the counterfeiter. They do find her, but she won't tell them who has the counterfeit gold that she made.

All the assistants learn from her is that one of the 12 bags contains counterfeit gold and that this bag's weight is different from the others. She refuses to reveal whether the different bag is heavier or lighter.

So the king needs to know two things.

• Which bag weighs a different amount from the rest?

• Is that bag heavier or lighter?

And, of course, he wants the answer found economically. He still has the old balance scale. He wants the solution in two weighings, but his court mathematician says it will take three weighings. No one else sees how it can be done in so few weighings. Can you figure it out?

Find a way to determine which bag is counterfeit and whether it weighs more or less than the others. Do so using the balance scale only three times.

My Dog's Smarter than Yours

Emiko claims her dog is very smart. When she opens the door in the morning, he runs out and brings in the newspaper.

You don't think that sounds very unusual? Well, Emiko lives in an apartment building. There are five newspapers outside every morning, and only one is a Japanese-language newspaper. That's the one that's delivered for Emiko's family.

Now, Emiko doesn't claim that her dog always brings in the right newspaper, but she says he does so more often than would happen if he were choosing randomly. Emiko's brother Hiro is skeptical. His hypothesis—the null hypothesis—is that the dog is simply picking papers at random.

Emiko and Hiro have devised a test. They will observe the dog for five straight days. Hiro says that if the dog brings in the right paper three or more times out of five, he'll reject his null hypothesis.

Suppose Hiro's null hypothesis is actually correct. What is the probability that he will end up rejecting it?

Defining Pascal

One way to define Pascal's triangle is by stating how each row is formed from the previous row. For example, the number 15 in the box here is the sum of the entries 5 and 10 just above it. In general, each entry in a new row can be found by adding the two closest entries in the previous row.

```
            1
          1   1
        1   2   1
      1   3   3   1
    1   4   6   4   1
  1   5  10  10   5   1
1   6  [15]  20  15   6   1
```

Another approach is to define Pascal's triangle in terms of the combinatorial coefficients. To do this, label the top row of Pascal's triangle as "row 0," the next row as "row 1," and so on. Similarly, label the entry at the left of any row as "entry 0," the next entry as "entry 1," and so on. For example, the boxed number 15 is entry 2 of row 6.

Using this system, we can define Pascal's triangle by this statement.

Entry r of row n of Pascal's triangle is the combinatorial coefficient $\binom{n}{r}$.

For example, entry 2 of row 6 should be $\binom{6}{2}$. Sure enough, $\binom{6}{2} = 15$. The entries 5 and 10 that add to give the entry 15 are the combinatorial coefficients $\binom{5}{1}$ and $\binom{5}{2}$.

It's important that these two ways of defining Pascal's triangle give the same result. In this activity, you will explore the relationship between the two definitions.

continued

1. Applying both definitions to this particular case of the entries 15, 5, and 10 gives this equation

$$\binom{6}{2} = \binom{5}{1} + \binom{5}{2}$$

You can verify numerically that $15 = 5 + 10$, but your task here is to explain this equation in terms of the meaning of combinatorial coefficients.

For example, why should the number of possible two-scoop bowls of ice cream, chosen from among six flavors, be the same as the sum of the number of one-scoop bowls chosen from among five flavors and the number of two-scoop bowls chosen from among five flavors?

2. Write a generalization of the equation displayed in Question 1. That is, write a general equation expressing the pattern for extending Pascal's triangle in terms of combinatorial coefficients.

3. Explain why your equation in Question 2 must be true for all values of n and r. Base your explanation on the meaning of $\binom{n}{r}$ as the number of ways to select r objects from a set of n objects.

Maximum in the Middle

One of the reasons for the interest in Pascal's triangle is that combinatorial coefficients play an important role in many probability problems.

For example, the combinatorial coefficient $\binom{6}{2}$, which is the number shown as $\boxed{15}$ below, gives the number of ways to flip a coin six times and get exactly two heads.

```
                    1
                 1     1
              1     2     1
           1     3     3     1
        1     4     6     4     1
     1     5    10    10     5     1
  1     6   [15]   20    15     6     1
```

If a coin is flipped six times, it makes sense that the most likely number of heads to get is three. This matches the fact that the greatest entry in the last row shown here for Pascal's triangle is 20. This entry corresponds to the combinatorial coefficient $\binom{6}{3}$. Remember that values for $\binom{n}{r}$ appear in what is actually the $(n + 1)$th row of Pascal's triangle, although we refer to this as "row n."

1. First consider the case in which n is even, so n is twice some other integer m. Prove that among all choices for r, the largest value of $\binom{2m}{r}$ occurs when r is equal to m. Base your proof on the formula for combinatorial coefficients.

2. State and prove a similar result for the case in which n is odd. You might begin by writing n as $2t + 1$ for some integer t.

The Whys of Binomial Expansion

You have expanded various expressions of the form $(a + b)^n$ and seen that the coefficients turn out to be combinatorial coefficients. In fact, this is true for every positive integer value of n. This principle is called the **binomial theorem.**

Your goal now is to find out why combinatorial coefficients appear in these expansions.

1. Use the distributive property to expand this expression.

$$(a_1 + b_1)(a_2 + b_2)(a_3 + b_3)(a_4 + b_4)$$

2. Each term in the expansion from Question 1 is a product of terms with some a's and some b's (or all a's or all b's). For instance, one of the terms is $a_1 b_2 b_3 a_4$, which has two factors that are a's and two factors that are b's.

    a. Altogether, how many terms in your expansion have two a's and two b's?

    b. What are the values of n and r for the combinatorial coefficient that best represents your answer from part a? Explain your answer.

3. Use your work from Questions 1 and 2 to write a general explanation of the fact that the coefficients of $(a + b)^n$ are combinatorial coefficients.

The Binomial Theorem and Row Sums

You have seen that the coefficients in the expansion of $(a + b)^n$ are the combinatorial coefficients that form row n of Pascal's triangle. This principle is called the *binomial theorem.*

In Pascal's triangle, the sum of the entries in row n is 2^n, as illustrated below. How can you prove this property using the binomial theorem? To begin, think about what can you substitute for a and b to make the expansion of $(a + b)^n$ equal to the sum of the entries from row n of Pascal's triangle.

$$1 = 1$$
$$1 + 1 = 2$$
$$1 + 2 + 1 = 4$$
$$1 + 3 + 3 + 1 = 8$$
$$1 + 4 + 6 + 4 + 1 = 16$$

PHOTOGRAPHIC CREDITS

Front Cover Photos

(From top left, clockwise) Jonathan Wong, Johnny Tran, Lindsey Macloud, Caroline Williams, Giovanni Guzman, Armani Wilson, Alida Jekabson, Eden Ogbai, Dylan Matthews, Thao Nguyen, Jenna Balch

Front Cover and Unit Opener Photography

Berkeley High School and Lincoln High School: Stephen Loewinsohn
Stock photos: iStockphoto

Pennant Fever

3 Lincoln High School, Stephen Loewinsohn; **5** iStockPhoto; **7** Klaus-Peter Wolf/Photolibrary; **10** Shutterstock; **11** Lincoln High School, Stephen Loewinsohn; **13** Shutterstock; **15** Tetra Images/Getty Images; **18** Silver Lake Regional High School, Kevin Sawyer, Lynne Alper; **22** Lori Adamski Peek/Getty Images; **24** iStockPhoto; **27** Lincoln High School, Lori Green; **38** SuperStock/SuperStock; **39** Hillary Turner; **42** iStockPhoto; **43** Shutterstock; **45** Bruce Ayres/Getty Images; **47** Lincoln High School, Stephen Loewinsohn; **55** Rayman/Photolibrary; **58** Creatas/Photolibrary; **59** iStockPhoto; **60** Kapa'a High School, Elaine Denny; **62** Photodisc; **66** PHOTOEDIT/PhotoEdit; **68** Wikipedia Public Domain; **69** Lincoln High School, Lori Green; **72** David Spindel/Superstock; **75** Lincoln High School, Stephen Loewinsohn; **77** Hillary Turner; **79** iStockPhoto; **86** iStockPhoto

www.ingramcontent.com/pod-product-compliance
Lightning Source LLC
Chambersburg PA
CBHW051227200326
41519CB00025B/7277